Disaster Zone

The Science Behind

HURRICANE HORRORS

Louise Spilsbury

CHERITON
CHILDREN'S BOOKS

Please visit our website, www.cheritonchildrensbooks.com to see more of our high-quality books.

First Edition

Published in 2022 by **Cheriton Children's Books**
PO Box 7258, Bridgnorth WV16 9ET, UK

Author: Louise Spilsbury
Designer: Paul Myerscough
Editor: Tim Cooke
Proofreader: Wendy Scavuzzo
Picture Researcher: Rachel Blount
Consultant: David Hawksett, BSc

Picture credits: Cover: Shutterstock/FotoKina; Inside: p1: Shutterstock/Arthur Villator; pp4-5: Shutterstock/FotoKina; p5b: Shutterstock/Multiverse; p6: PXHERE.COM; p7: Shutterstock/MarcelClemens; p8: Shutterstock Arthur Villator; p9: Shutterstock/Terry Kelly; pp10-11: Shutterstock/AMFPhotography; p11r: Shutterstock/MDay Photography; pp12-13: Shutterstock/Drew McArthur; p13r: Wikimedia Commons/Presidencia de la República Mexicana; p14: Shutterstock/NASA images; p15: Wikimedia Commons/UW/CIMSS/William Straka III; p16: Shutterstock/Michael Gerrard; pp18-19: Shutterstock/Ingridagrants; p19b: Wikimedia Commons/Christian Frausto Bernal; pp20-21: Shutterstock/Ymphotos; pp20-21b: Wikimedia Commons/NASA/Nilfanion; p22: Shutterstock/Terry Kelly; p23: Shutterstock/Felix Mizioznikov; pp24-25: Wikimedia Commons/U.S. Customs and Border Protection photo by Kris Grogan; p25b: Wikimedia Commons/Roosevelt Skerrit from Vieille Case, Dominica; pp26-27: Wikimedia Commons/United States Air Force/Tech. Sgt. James Pritchett; p27b: NOAA/U.S. Air Force photo by Senior Airman Anna-Marie Wyant; p28: Wikimedia Commons/NASA/NOAA GOES Project; p29: Wikimedia Commons/NASA; p30: U.S. Air Force photo by Master Sgt. Jessica Kendziorek; p31: Wikimedia Commons/U.S. Air National Guard photo by Staff Sgt. Daniel J. Martinez; p32-33: Wikimedia Commons/U.S. Coast Guard, Petty Officer 2nd Class Kyle Niemi; p33r: Shutterstock/Tad Denson; pp34-35: Shutterstock/Kekyalyaynen; p35r: NASA/Lauren Harnett; p36: Shutterstock/Forestpath; p37: Wikimedia Commons/Cyclonebiskit; pp38-39: Shutterstock/Ronnie Chua; p39b: Wikimedia Commons/David Shankbone; pp40-41: Shutterstock/Jesse Stephens; pp40-41b: Shutterstock/Cire notrevo; p42: Shutterstock/FrameStockFootages; p43: Wikimedia Commons/Ed Edahl/FEMA; pp44-45: Shutterstock/Alan Budman; p45b: Wikimedia Commons/U.S. Air Force photo/Staff Sgt. Kelly Goonan; p47: Shutetrstock/NASA images.

Printed in the United States of America

Contents

Chapter 1

WILD AND DANGEROUS WINDS

Hurricanes are the deadliest and most powerful storms on Earth. These swirling, fast-moving winds bring heavy rain and can cause terrible destruction over vast areas.

Hurricane Horrors

Hurricanes form over warm ocean waters. Some are not very big or fast, and will stay far out to sea. They weaken and fizzle out before anyone sees or feels them. Other, bigger hurricanes move quickly toward land. They can hit shorelines at high speed and with powerful force. Hurricanes usually cause the most damage at the coast, which is where they first arrive. The deadliest hurricanes can rage on and cause destruction farther inland, too. Some hurricanes last only for a few hours. The worst ones can last for days or even weeks. They can injure and kill thousands of people.

Naming Hurricanes

Hurricanes occur every year. Sometimes, two or three hurricanes happen at the same time. Giving each hurricane a name makes it easier to keep track of the different storms—everyone knows which storm is being talked about. It means that weather scientists, hurricane researchers, emergency response workers, ship captains, and other people can all talk about the hurricane and know it's the same one. Hurricanes can have male and female names and the names can be repeated after six years. The names are always in alphabetical order. Scientists do not use the letters Q, U, X, Y, or Z. That's because few names begin with those letters.

DEADLY NAMES

The names of some of the worst hurricanes are used only once. For example, in 2017, hurricanes Harvey, Irma, Maria, and Nate caused terrible damage and loss of life. Out of respect for the people who suffered in the storms, the hurricane names will not be used again for future hurricanes.

This street **flooded** after a catastrophic hurricane hit Fort Lauderdale, Florida, in 2017.

The hurricane that caused the destruction shown in this photo was named Hurricane Irma.

DEADLY DISASTERS

The word hurricane comes from the name for the Carib god of evil, Hurican. The Maya also had a god of storms and floods named Hurakan. In the Caribbean, violent storms were named hurricane because of their destructive power.

Whirling Winds

Hurricane wind speeds range from 74 miles per hour (119 kph) to more than 150 miles per hour (240 kph). The strongest hurricane winds can reach 200 miles per hour (320 kph). That is as quick as the fastest trains on Earth. No one stands a chance of outrunning these monsters.

Damage to Land

Hurricanes that hit land can flatten buildings and uproot trees. They can throw and spin large objects around in the air like they were small toys. They can smash windows, blow roofs off buildings, and hurl mobile homes down streets. As hurricanes pick up branches and other objects, that **debris** adds to the dangers. People on the streets trying to take cover can be blown over and risk being hit by flying debris.

When Hurricane Dennis reached the coast of Florida in 2005, it brought winds of 120 miles per hour (193 kph).

HAIL HAVOC

Hail is rare in a hurricane. Usually, the hurricane's warm center melts hail before it reaches the ground, but hail does happen sometimes. Hailstones form in high hurricane clouds as water freezes into layers over ice particles, or pieces. When hailstones smash to the ground, they can damage cars and other objects. Hail can badly injure people, especially when thousands of hailstones drop to the ground all at once.

Large hailstones, like the one shown here, can cause injuries if they hit a person.

Q What type of flying debris might harm people the most during a hurricane?

Fearsome Floods

Hurricanes drop huge amounts of water quickly. The rain does not fall gently or softly, either. Hurricane winds are fast, and it can feel like they are firing raindrops against people and objects. The heavy rains that hurricanes bring can pour down suddenly. That can cause floods. Land that is normally dry, such as streets and fields, gets covered by large amounts of water in a short time. Heavy and fast-moving flood water can knock people off their feet and drown them. The rains swell rivers, too. They then run so fast, they can wash away bridges, parts of roads, cars, and people.

A TRAIL OF DESTRUCTION

Large hurricanes are so dangerous they can cause terrible damage. They will wreck almost all buildings they hit, and can injure and kill large numbers of people. Some of the most violent hurricanes also set off other deadly natural disasters.

Storm Surge

Storm surges are the number-one killer in hurricanes. A storm surge is a bulge of water caused when winds whip up water in front of a hurricane. These surges cannot really be seen out in the open ocean where the water is deep. As a hurricane nears a coastline where the water becomes shallower, the bulge of water starts to grow. When the storm surge hits land, the water can be 20 feet (6 m) high and many miles wide. Storm surges move fast and with great force. They can pick up boats, cars, and homes as they move onto land. Objects picked up by the storm surge can be carried far out to sea. After the hurricane has moved away water drains back into the ocean. Storm surges wreck anything in their path and leave a trail of destruction behind.

When a storm surge crashes over a coastline, it can cause massive flooding, destroying everything at ground level.

Landslide!

Heavy rains from a hurricane can also soak soil-covered hillsides and slopes. This can make the soil heavy and loose. When the muddy soil and rocks become too heavy to stay on the slope, they can suddenly slide downhill very fast. When they get to the bottom of the slope, they can bury anything in their path, including entire towns. That is called a landslide.

In 2018, Hurricane Michael destroyed just about everything with a massive storm surge, intense winds, and landslides when it made **landfall** at Mexico Beach, Florida.

Where in the World?

In September 2018, Hurricane Florence brought devastating flooding and damaging winds. In parts of North Carolina, there were reports of a storm surge. It was 10 feet (3 m) above normal water levels. Residents had followed warnings and moved to safety. However, the storm surge caused a lot of damage to coastal homes.

Disaster Zone: Texas

Hurricane Harvey struck Texas on August 25, 2017. It hung over Texas for four days and caused a lot of damage. Its wind speeds of 130 miles per hour (209 kph) and heavy rains caused terrible **flash floods** and river flooding. A massive storm surge **submerged** coastal areas—the storm surge alone flattened buildings and destroyed entire blocks.

This overhead view shows the extent of the flooding caused by Hurricane Harvey in 2017 in Houston, Texas.

SPIRALING TWISTERS

Hurricane Harvey also created tornadoes, or twisters. Tornadoes often form in the bands of rain outside the hurricane. These destructive spirals of air can spin off in all directions. Harvey may have created more than 50 tornadoes. Hurricane Harvey caused about $125 billion in damage. It is the second most costly hurricane to hit the United States mainland since 1900.

Sinking a City

Harvey was one of the wettest storms in US history. By the time it blew itself out, it had dropped more than 27 trillion gallons (102 trillion l) of rain over Texas. In some parts of the city of Houston, more than 50 inches (127 cm) of rain fell. One-third of the city was flooded. So much water fell that experts believe it made the city sink by almost 1 inch (2.5 cm) for a short time.

Flooded Homes and Streets

At least 30,000 people had to move to shelters because their homes were filled with water. Their belongings were washed away or ruined. Rescue workers across Texas worked night and day to reach stranded residents and move them to safety. Tens of thousands of people who stayed in their homes were left with no electricity or clean water. A **dam** burst and there was a chemical spill from a pipeline that broke. People were told not to touch floating fire ants, which had been washed from their underground homes, because they sting and bite.

Some of the twisters formed by Hurricane Harvey ripped apart homes and other buildings.

Q

Why do you think a lack of clean water is a problem after a hurricane?

Chapter 2

HOW HURRICANE HORRORS WORK

All hurricanes, large or small, form over areas of warm ocean water. They take heat from the ocean, then convert that heat into the energy of their winds. The warmer and wetter the air is above an ocean, the larger and deadlier the hurricane.

A Hurricane Is Born

A hurricane starts to form when wind blows over the surface of a warm ocean. The wind takes some of the warm water from the surface and turns the liquid water into **water vapor**. The warm water vapor rises. It is lighter than the colder air above. When the water vapor rises from the ocean, there is less air left near the ocean's surface. Cooler air pushes in to fill the gap left by the warm, rising water vapor. Then that air becomes warm and moist, and it rises, too, creating a plume of rising air. That is called an updraft.

Becoming a Hurricane

A hurricane grows by sucking up more and more air. As warm air rises higher into the **atmosphere**, the water vapor cools and changes. It turns into drops of liquid water. The water droplets form a column of cloud that gets higher and larger. As more air is added, it starts to spin in circles around the center. At the middle, the air speeds up. It spins faster and faster, like an ice skater when they pull in their arms and legs. When the spinning winds reach speeds of 39 miles per hour (63 kph) or more, they become a **tropical** storm. By 74 miles per hour (119 kph), the storm has become a hurricane. As hurricanes blow away from warm areas of water and toward cooler areas of land, the water vapor turns to liquid water. That falls as rain.

HOTTER AND STRONGER

Water on the surface of an ocean must be 79 degrees Fahrenheit (26 °C) or higher to create the energy a hurricane needs to form. A few degrees in temperature can make a big difference to the hurricane's strength. Hurricane Patricia, which blasted Mexico in 2015, was one of the most powerful hurricanes ever recorded. It developed over water at 86 degrees Fahrenheit (30 °C). It caused a lot of damage. Around 350 trees were ripped out of the ground in Colima (shown right). Landslides blocked the Colima-Manzanillo highway.

Trees ripped from the ground by Patricia were then dropped back on the land, forming piles of debris.

Warm, moist air and wind are needed for a hurricane to form.

THE EYE OF A HURRICANE

Hurricane winds rotate, or spin, in a spiral. They form a tall column of clouds that spin fiercely around a central point. That central point is called the eye of the hurricane. In the eye, it is very calm and strangely still.

Looking into the Eye

In the eye of a hurricane, there are no winds or clouds. There is little rain, because it is there that the air sinks downward. A hurricane's eye can be huge—in bigger hurricanes, it can be 40 miles (64 km) wide. If you are caught in a hurricane, the winds and heavy rain suddenly stop and the sky clears when the eye moves over you. Then, just as quickly, the winds and rain begin again, but this time moving from the opposite direction.

The Eye Wall

The eye wall is the name of the high funnel-shaped column of wind and thunderstorm clouds that are just around the edge of the hurricane's eye. It is a thick ring of vertical clouds. The eye wall is the most intense and dangerous part of a hurricane. It is where a hurricane's most damaging winds and heaviest rainfall are found. In some hurricanes, those winds can reach staggering speeds of 140 miles per hour (225 kph).

This picture of Hurricane Florence shows its swirling winds and central eye.

Far-Reaching Danger

A hurricane's deadly powers do not end with the eye wall. The outer parts of hurricanes contain strong winds and clouds that can dump heavy rains. These bands of clouds and rain can stretch for hundreds of miles. They can cause a lot of trouble. They drive the push of ocean water that can create storm surges. When hurricanes travel across warm land, it is these outer winds that can form tornadoes.

The largest hurricanes can be more than 600 miles (966 km) wide, and there can still be windy and stormy areas beyond that range, too. This **thermal image** of Hurricane Patricia was taken by a **satellite** in space.

Where in the World?

In 2003, Hurricane Isabel hit the eastern United States. It made an impact as far as 115 miles (185 km) from its eye. The eye itself was 40 to 50 miles (65 to 80 km) wide. The high winds and heavy rain battered parts of North Carolina; Virginia; Washington, DC; Maryland; and Delaware. At least 35 people were killed and millions more people were left without power.

HURRICANE HORRORS ON THE MOVE

Hurricanes form and grow into monsters over the oceans. However, they do not stay there. They can move, and change direction. They can race toward land, where they can cause terrible damage.

Fast and Furious

Hurricanes move like a spinning top. While they are spinning quickly around a central point, the winds in the atmosphere cause hurricanes to move in one direction. Hurricanes can travel thousands of miles over oceans, and they feed off the energy of warm seas.

Hurricane Isabel was a Category 5 hurricane that caused immense destruction as it moved across Virginia in 2003.

How Hurricanes Die

Hurricanes need two things to form: warm water and regular winds. The winds cannot change greatly in speed or direction before they spiral into the sky. When a hurricane reaches land, there is no more water. Over land, the winds may die down, too. They are slowed down by objects such as trees, buildings, and hills. When hurricanes do not have the forces that gave them energy, they start to slow down and die, like a spinning top slowly winding down. Large hurricanes do not run out of energy until it is too late. By then, they have caused terrible destruction over coastal land and communities.

Q

Can you think of objects in the natural world that could help slow down a hurricane?

HURRICANE SCALES

The Saffir-Simpson Hurricane Wind Scale ranks hurricanes on a scale from one to five. It is based on their greatest wind speed and the possible damage they could cause.

Category 1: Winds of 74 to 95 miles per hour (119 to 153 kph). These winds can damage roofs, shingles, gutters, power lines, and poles. They can snap large tree branches, and topple some trees.

Category 2: Winds of 96 to 110 miles per hour (154 to 177 kph). They will damage mobile homes and roofs. They blow down trees, and cause power outages.

Category 3: Winds of 111 to 129 miles per hour (178 to 208 kph). They cause more serious damage to buildings. Mobile homes and larger trees will be destroyed. Flooding inland will take place.

Category 4: Winds of 130 to 156 miles per hour (209 to 251 kph). These winds will do enough damage to make an area unlivable for months. Extensive flooding will occur.

Category 5: Winds of more than 157 miles per hour (252 kph) or higher. These cause widespread catastrophic damage and flooding. They destroy most buildings and life in their path.

Disaster Zone

Mexico, Central America

Most hurricanes take days to form. However, in 2015, the eastern Pacific Ocean was warmer than usual. The temperature was more than 86 degrees Fahrenheit (30 °C) and this was ideal for hurricanes to form. In a single day, one storm transformed. It went from a tropical storm to a Category 5 hurricane. Hurricane Patricia became the strongest hurricane to reach Mexico in more than 50 years.

Powerful Patricia

Hurricane Patricia blasted the coast with winds of more than 200 miles per hour (320 kph). It was one of the strongest hurricanes ever recorded. Strong winds tore roofs off houses. They pulled up, snapped, and stripped nearly all trees. Hillsides were left with no trees. Power poles and **transmission towers** were crumpled by the winds. Heavy rains caused widespread flooding, and houses were flooded as rivers burst their banks.

Patricia damaged or destroyed more than 10,000 homes. Around 100,000 acres (40,470 ha) of farmland was destroyed. Plantain, banana, and papaya **crops** were wiped out. Mexican authorities had warned all towns and villages in Patricia's path to leave. Thousands of tourists and local people had left the danger zone and moved into shelters. Airports were closed. Homes and store fronts were boarded up for protection. Two people were killed when winds blew a tree onto their tent at a campsite and four more people died in a car accident.

When Patricia struck, it destroyed beach cafés before moving inland to cause further damage.

A WEAKER MONSTER

Patricia was a monster hurricane, but it caused less damage than other big hurricanes. That is because, when the hurricane made landfall along Mexico's Pacific coast, it hit the 10,000-foot (3,050 m) and higher mountains in the Sierra Madre range. The collision began to weaken the storm before its eye wall came ashore. Also, Patricia's most dangerous core of winds was fairly narrow. It passed between two big cities, Puerto Vallarta and Manzanillo, instead of hitting them.

The Sierra Madre mountain range is in Mexico.

Chapter 3

WHERE HURRICANE HORRORS HAPPEN

Hurricanes always start over tropical oceans. They start in parts of the world around the **equator**, such as the northwest Pacific, west Atlantic, and Indian Oceans. Once a hurricane develops, it can travel long distances to the nearest coast. Most hurricanes that hit the United States begin in either the Caribbean Sea or the Atlantic Ocean.

The wind storm that caused this destruction in the Philippines was named Typhoon Haiyan, and was one of the strongest typhoons ever recorded.

Different Places, Different Names

Storm systems have different names. Hurricanes only happen over the Atlantic and northeast Pacific Oceans. Storm systems that occur in the northwest Pacific are called typhoons. They are called cyclones if they start in the south Pacific and Indian Oceans. Even though they have different names, they are the same kinds of storm. They form in the same way, and they can cause similar damage and destruction. The only real difference is that they spin in different directions, depending on where in the world they start.

Spin Directions

Cyclones that begin in the southern **hemisphere** spin **clockwise**. Hurricane winds that start in the northern hemisphere spin **counterclockwise**. This happens because planet Earth spins slowly on its **axis** all the time. It makes one complete turn every 24 hours. As Earth spins, it sweeps the winds blowing above its surface in different directions. In the southern hemisphere, winds curve to the left. In the northern hemisphere, as air travels north, winds curve right.

This map shows the tracks followed by hurricanes as Earth spins and drags them around the world.

MAKING LANDFALL

Where a hurricane hits land makes all the difference because its impact changes. If a hurricane hits a coast and travels onto the land, this is a direct landfall. This means the worst damage of the hurricane will be in one area of land. If a hurricane moves **parallel** to a coastline, its effects are spread over a much larger area.

HURRICANE SEASONS

Hurricanes do not happen all year round—they only occur in certain seasons. They usually strike during the warmer months of the year, which are known as hurricane seasons. Anyone who lives in or visits areas where hurricanes happen needs to be prepared for a hurricane event.

Atlantic Highs

The official hurricane season for the Atlantic Basin is between June 1 and November 30. The Atlantic Basin includes the Atlantic Ocean, the Caribbean Sea, and the Gulf of Mexico. The peak of the season, when the worst hurricanes happen, is from mid-August to late October. That is partly because wind shear is at its weakest at that time. Wind shear is the name for a sudden change in wind speed and/or direction over a short distance. When there is a lot of wind shear, hurricanes cannot form. When there is less wind shear, hurricane winds can spin upward more easily. Also, it's much hotter from mid-August on, and sea temperatures are higher. Warmer air above the oceans means a hurricane is more likely.

Hurricane Michael occurred in October, during the hurricane season of 2018.

Where in the World?

The five cities in the United States where hurricanes are most likely to occur are Miami, Florida; Key West, Florida; Cape Hatteras, North Carolina; Tampa, Florida; and New Orleans, Louisiana. Florida has been hit by more hurricanes than any other state. Residents of Florida know this and get ready for hurricane season.

People board up their windows and make their buildings safe to prepare for a hurricane.

Survey of the Seasons

Northwest Pacific Ocean hurricanes start in July and can go on until November. In the northeast Pacific Ocean, violent storms start as early as May and there is still some risk of hurricanes until November. When one hurricane season ends in one part of the world, it is just starting in another. In the southwest Pacific and the south Indian Oceans, hurricanes can form from October to May. The north Indian Ocean has the longest hurricane season. There hurricanes can happen on the coasts any time between April and December.

Q

How do you think understanding hurricane seasons can help people in hurricane hotspots?

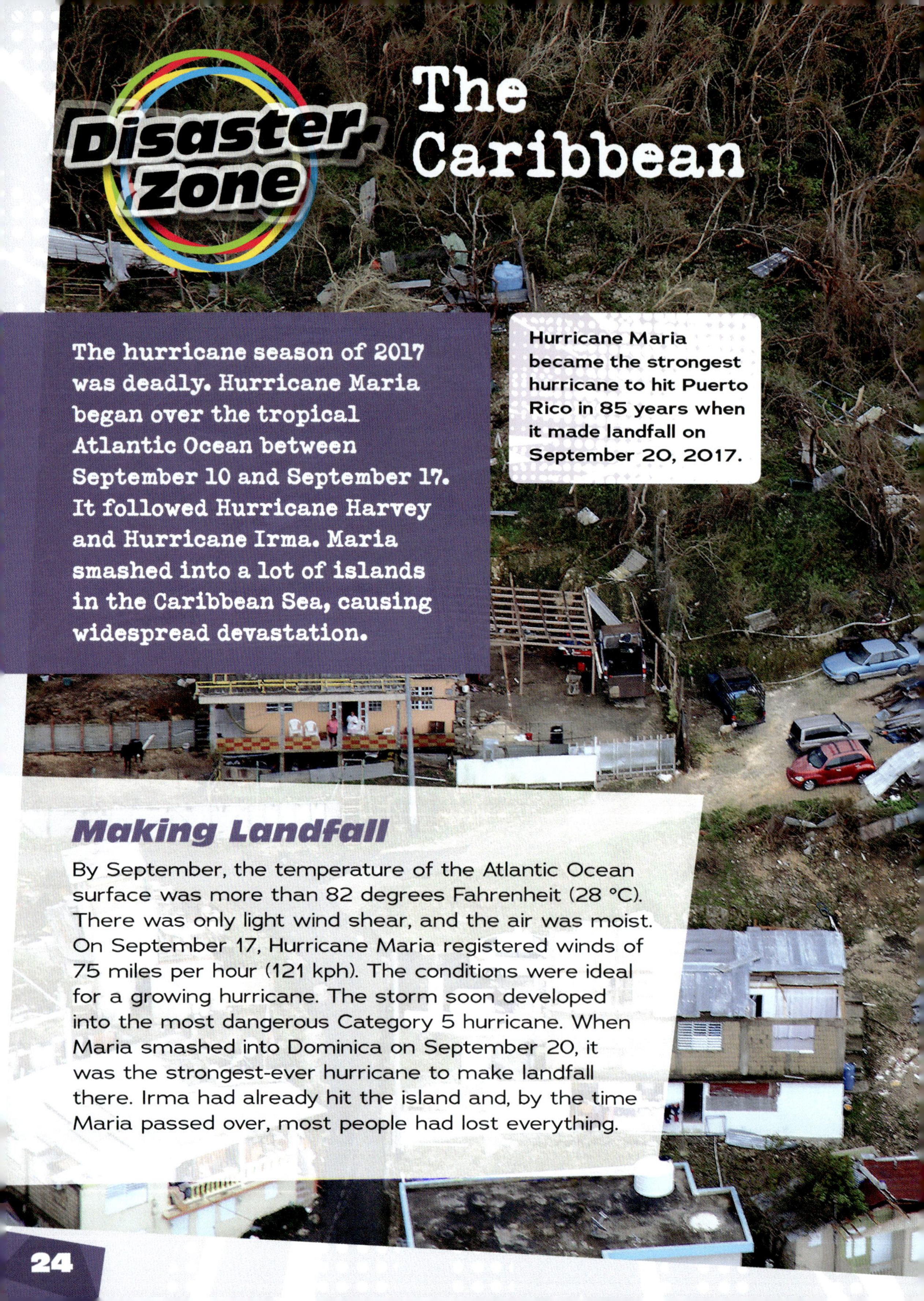

Disaster Zone

The Caribbean

The hurricane season of 2017 was deadly. Hurricane Maria began over the tropical Atlantic Ocean between September 10 and September 17. It followed Hurricane Harvey and Hurricane Irma. Maria smashed into a lot of islands in the Caribbean Sea, causing widespread devastation.

Hurricane Maria became the strongest hurricane to hit Puerto Rico in 85 years when it made landfall on September 20, 2017.

Making Landfall

By September, the temperature of the Atlantic Ocean surface was more than 82 degrees Fahrenheit (28 °C). There was only light wind shear, and the air was moist. On September 17, Hurricane Maria registered winds of 75 miles per hour (121 kph). The conditions were ideal for a growing hurricane. The storm soon developed into the most dangerous Category 5 hurricane. When Maria smashed into Dominica on September 20, it was the strongest-ever hurricane to make landfall there. Irma had already hit the island and, by the time Maria passed over, most people had lost everything.

Maria on the Move

As Maria moved on, the hurricane mostly missed the island of Martinique. It caused power outages across Guadeloupe. When Maria hit Puerto Rico, it was still a Category 4 hurricane. The hurricane blasted the region with winds of 155 miles per hour (249 kph). Palm trees and power lines were snapped in half, and the entire island lost power. Homes, roads, and bridges were destroyed, and extremely heavy rainfall caused major flooding. The northern half of Puerto Rico was hit the worst.

The Impact on Puerto Rico

The effects of a hurricane like Maria carry on after the storm is over. There were food and water shortages. Toilets could not flush and there was no water for showers or washing clothes. There were outbreaks of diseases caused by **polluted** flood waters, and the lack of electricity went on for a long time. Important buildings such as hospitals, schools, and banks were badly damaged. Life was difficult for months after the hurricane.

KILLER HURRICANE

Maria killed 64 people. They drowned, were crushed by collapsing buildings, or hit by flying debris. In the six months after the hurricane, many more died. This was because of poor health care, repeated power outages, and a lack of electricity and clean water. One year later, the final death toll was more than 3,000.

It can take people a long time to rebuild their homes and lives after the massive destruction left in the wake of a hurricane like Maria.

Chapter 4

SCIENTISTS ON THE CASE

Scientists called meteorologists study the weather. They use a number of ways to spot and study hurricanes. This helps them better understand what makes these storms happen. It also helps them **forecast** where and when a hurricane might hit. They can then warn people to get out of the path of the storm.

Hurricane Hunters

Hurricane hunters are scientists who chase hurricanes and fly into the storms to collect information. They study the size, strength, speed, and direction of a hurricane. They also study the possible landfall of the storm. To figure out where rain is falling, the airplanes used by hurricane hunters carry special scientific equipment. The airplane's **radar** systems send out radio waves, which record signals as they bounce off clouds and rain. The radar systems also help scientists figure out how hurricane storm clouds form. Other onboard devices can measure ocean surface conditions. That can be very helpful for **predicting** storm surges.

Expert pilots fly hurricane-hunting aircraft through the violent winds and rain of a hurricane so scientists can get the best results from their equipment.

One of the most important tools used by hurricane hunters is the dropsonde. A dropsonde is a lightweight cylinder that is dropped by a parachute into a storm. Dropsondes contain **sensors**. As they fall toward the ocean, they send out readings. They collect information on wind direction and speed, **air pressure**, temperature, and **humidity** from positions throughout the hurricane. They have **Global Positioning System (GPS)** technology so they can **transmit** their position every half second as they drop through a storm toward the ocean. On one flight, scientists might send 20 dropsondes into a storm in various spots. They cover from the eye to the outer rim.

LIFE AS A HURRICANE HUNTER

Hurricane hunters' airplanes are designed to take the force of a hurricane storm. Some of them can fly directly into the worst of the hurricane and through the eye. Scientists can be in the air for 12 hours at a time, and it can be very bumpy! Some scientists describe it as being on a scary roller-coaster ride. It is very noisy, too, because rain beats down on the aircraft.

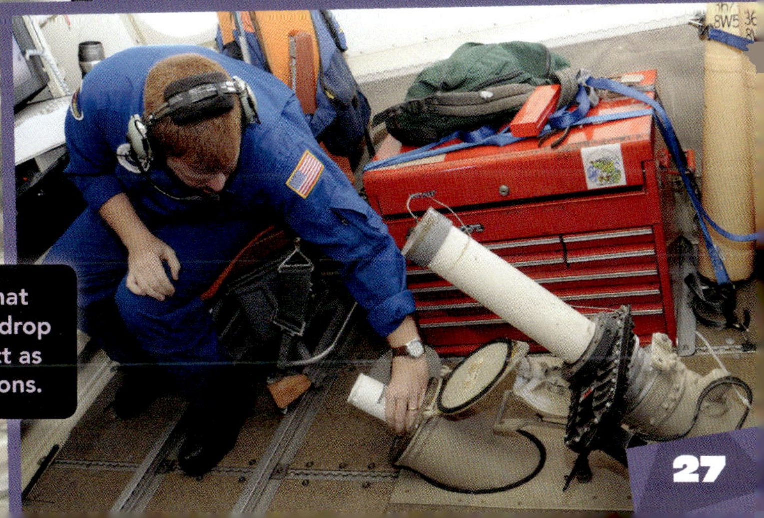

The dropsondes that hurricane hunters drop into a hurricane act as mini weather stations.

SATELLITE STUDIES

To get a more complete picture of hurricanes and their exact routes, it helps to look at Earth from space. Scientists do this by using pictures of clouds taken by weather satellites, which sit high above Earth. They help scientists figure out which clouds might become hurricanes.

An Eye in the Sky

A satellite is an object that moves around a planet, star, or moon. The moon is a natural satellite of Earth. Satellites that people launch into space are human-made satellites. They are machines that are placed high above Earth's atmosphere and are in **orbit** around the planet. Geostationary satellites move around Earth at the same rate that Earth rotates. That means the satellite is in the same place at the same time every day. Polar-orbiting satellites circle the planet in a route that takes them near Earth's poles.

This satellite image shows the type of large clouds that can develop into severe storms and hurricanes.

Watching the Water

Weather satellites have cameras that take pictures of oceans. They help people spot new storms and record how they are growing. They also carry cameras that can detect the different temperatures of clouds. At the ocean's surface, they record the temperature. Scientists can figure out when waters are warm enough to create a hurricane. Some cameras have sensors that measure rainfall patterns, and the shape of the hurricane through the clouds.

Seeing into Storms

One very useful satellite takes pictures of the inside of hurricanes. It has a radar system that can see inside storms. It gives scientists a three-dimensional (3D) view beneath the clouds, which helps them identify which storms have hot towers. Hot towers are rain clouds that reach the top of the troposphere. That is the lowest layer of Earth's atmosphere. By studying a lot of hurricanes, scientists know that storms with hot towers are more likely to grow bigger over 24 hours and turn into monster hurricanes. That is because the hot towers carry heat high up into a hurricane.

TRACKING A MONSTER

Polar satellites are usually placed into a low orbit around Earth. They sit between 60 to 900 miles (97 to 1,448 km) above Earth's surface. A polar satellite will orbit Earth once every 90 minutes. Each orbit takes it on a slightly different route above Earth's surface as the planet rotates beneath it.

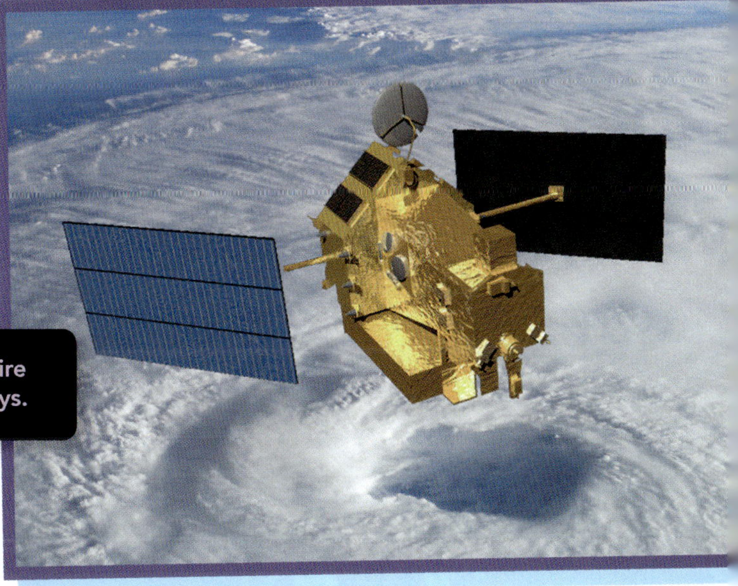

A polar satellite can view the entire surface of Earth in one to two days.

FORECASTING HURRICANES

Meteorologists are always working on ways to study hurricanes. This helps them better forecast and predict what the hurricanes will do and how they will grow.

Drones at Work

Hurricane hunting is an important job, but it is hard work and can be dangerous. That is why scientists prefer to use drones, which are unmanned aircraft that can fly inside hurricanes. Drones are controlled by handsets from miles away, either on the ground or onboard an aircraft. A huge advantage of drones is that they can travel high into a hurricane to collect **data** where hurricane hunters cannot reach. Drones can be equipped with specialized sensing equipment that measures wind, rain, and temperature. It can measure the moisture in the air, and the air pressure above and inside the storm. Drones can even release dropsondes above storms.

Thanks to more powerful computers that help meteorologists prepare better storm models, scientists are able to release forecasts of hurricanes up to five days before the storms make landfall.

Supercomputers

Meteorologists use very powerful computers called supercomputers, which analyze billions of pieces of weather information. This information is collected from drones and different sources around the world. Meteorologists take all the information and use it to create computer forecast models. These models are like pretend storms, and they allow scientists to forecast a hurricane's path. They can predict any changes in the hurricane before it hits land. All this information helps governments and emergency services prepare for the approaching storm. No forecast can be exact, however, because there are so many things that can change. For example, unexpected warm currents of water in oceans can strengthen a hurricane. Even the best data could quickly be outdated.

INSPECTING THE DAMAGE

Drones can also be used in any rescue operation during or after a hurricane. They can fly over a wide area to give an overview of the damage. They can spot people in need of help, and they can also deliver rescue ropes and life jackets in areas that are too dangerous for ground-based rescuers. After Hurricane Harvey in 2017, dozens of drones were used in Houston, Texas. They inspected roadways, checked railroad tracks and power lines, and also inspected the damage at the coastlines.

This photo of the floodwaters created by Hurricane Harvey was taken by a drone.

Q If you were a meteorologist, what important information would you look for to predict a hurricane?

Disaster Zone

Louisiana

Scientists work hard to forecast where a major storm is heading and how big it will be. However, hurricanes are wild storms and cannot always be predicted. In 2005, Hurricane Katrina surprised experts with catastrophic results.

In some places in New Orleans, floodwaters were 20 feet (6 m) deep.

A Surprising Storm Surge

Hurricane Katrina formed over the Caribbean Sea on August 23, 2005. It made landfall in Florida on August 25, where it killed nine people. Then it moved to the Gulf of Mexico, where it suddenly changed, becoming much wider. It set off an unexpected and terrible storm surge into New Orleans. When it hit the city on August 29, its wind speeds reached 125 miles per hour (200 kph). The storm also brought very heavy rain. The storm surge that hit New Orleans was almost 30 feet (9 m) high and was the biggest ever recorded on the coast of the United States.

Damage and Devastation

Water is heavy. When it slams quickly into objects, it can demolish anything in its path. The city of New Orleans lies below sea level, so it had banks of earth called levees built around it to protect the city against flooding. However, the storm surge smashed through the levees and water rushed into the city. About 80 percent of the city was flooded. The water rushed through the streets and poured into people's homes. It carried poisonous snakes, dirt, rocks, and debris from damaged buildings all over the city. Most of the people who were killed during Hurricane Katrina died due to the storm surge.

Warnings were given the day before Katrina hit New Orleans. People had begun to evacuate, but many were still in the city when it flooded.

DEATH AND DESTRUCTION

Katrina's winds snapped trees and tore out windows. They tossed furniture from homes onto the streets, and broke power lines. The storm moved along the coast, causing more damage in Louisiana, Mississippi, and Alabama. Katrina tossed boats ashore and filled towns with mud and sand. In the town of Biloxi, Mississippi, a storm surge 30 feet (9 m) high killed 30 people. In total, Katrina caused destruction up to 150 miles (241 km) inland. More than 500,000 people were left homeless and 1,850 people died. It was the deadliest hurricane to hit the United States in 75 years.

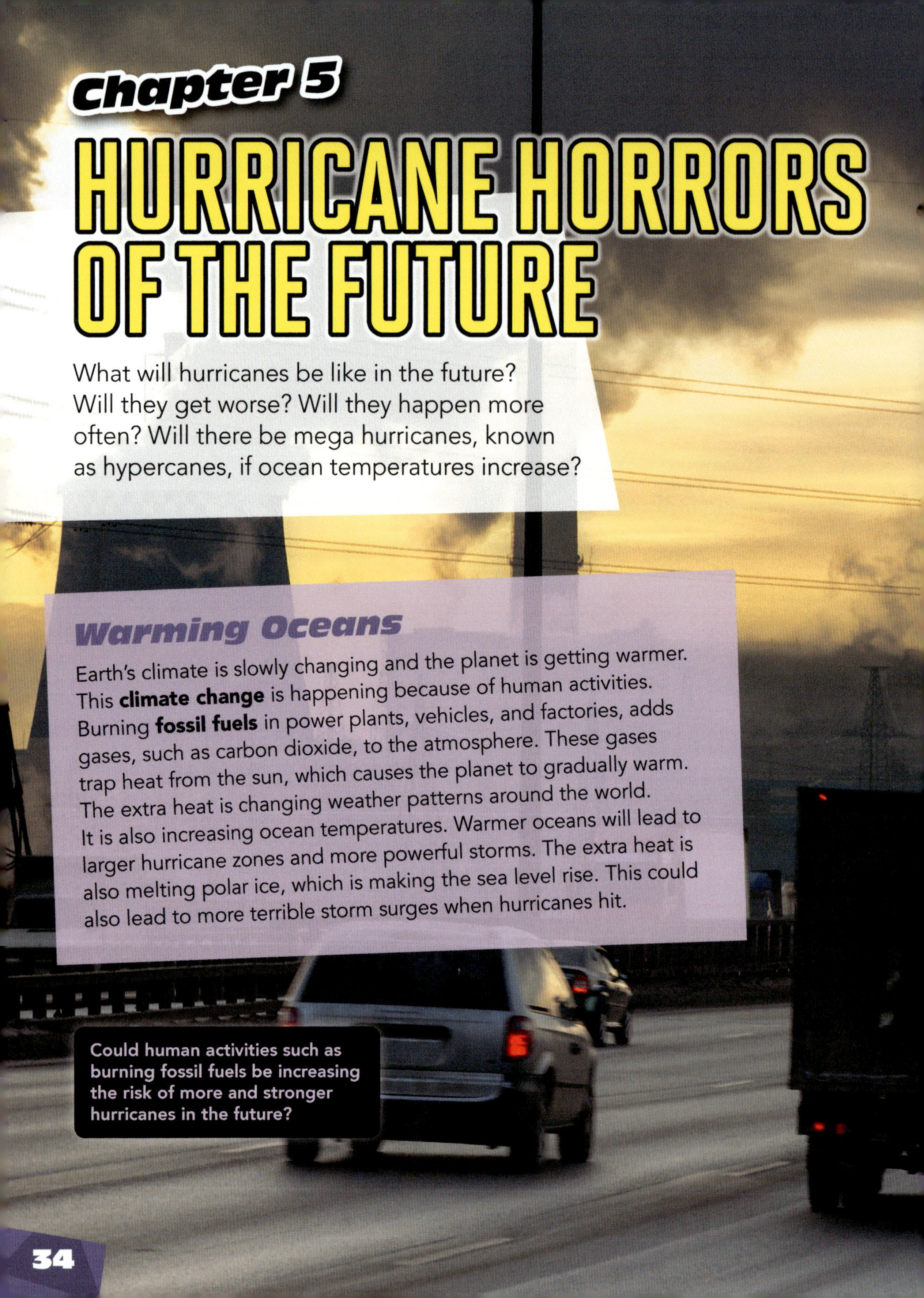

Chapter 5

HURRICANE HORRORS OF THE FUTURE

What will hurricanes be like in the future? Will they get worse? Will they happen more often? Will there be mega hurricanes, known as hypercanes, if ocean temperatures increase?

Warming Oceans

Earth's climate is slowly changing and the planet is getting warmer. This **climate change** is happening because of human activities. Burning **fossil fuels** in power plants, vehicles, and factories, adds gases, such as carbon dioxide, to the atmosphere. These gases trap heat from the sun, which causes the planet to gradually warm. The extra heat is changing weather patterns around the world. It is also increasing ocean temperatures. Warmer oceans will lead to larger hurricane zones and more powerful storms. The extra heat is also melting polar ice, which is making the sea level rise. This could also lead to more terrible storm surges when hurricanes hit.

Could human activities such as burning fossil fuels be increasing the risk of more and stronger hurricanes in the future?

Learning from the Past

To learn what hurricanes of the future might be like, scientists study past hurricanes. For example, in Earth's distant past, a monster hurricane sent a storm surge washing over land and into lakes. It left behind a fan-shaped zone of sand. Over time, this layer of sand was buried by more layers of soil. Scientists can drill down for samples of the ancient sand and figure out how old it is. That can tell them when the giant storm happened. They can tell how big the hurricane was by how far inland the storm surge spread. They compare that to what they know about Earth's temperature at different times in the past. Then they can see how hurricanes get bigger or smaller as climates change.

Scientists onboard an aircraft investigate how a combination of storms and air pollution from wildfires, cities, and other sources can change our climate.

HYPERCANE HORRORS

A hypercane is the deadliest hurricane of all. It is an extreme hurricane that experts think could form if ocean temperatures reach 122 degrees Fahrenheit (50 °C). A hypercane could be 25 miles (40 km) high. It would stretch all the way into the upper atmosphere. Its winds would be so loud they could almost break the **sound barrier**. Hypercanes are believed to have all but wiped out life on the planet 66 million years ago. If one forms again, it could destroy life on Earth.

MODELING HURRICANE HORRORS

Scientists are developing better tools to help them understand hurricanes. They want to make forecasts more accurate. There is a race to make this happen quickly because hurricanes are getting bigger and more dangerous.

Category 6?

Category 5 hurricanes are the highest on the Saffir-Simpson Scale. They can have winds of 157 miles per hour (252 kph) or more. A Category 5 hurricane will destroy cities such as New Orleans. Today, a Category 6 hurricane does not exist. However, experts now believe there soon will be one. A Category 6 hurricane's winds would be more than 200 miles per hour (322 kph). Some experts think it is only a matter of time before a devastating Category 6 storm hits the United States.

This photo shows Hurricane Ike hitting the coast of Florida. The storm surge battered everything in its path. Ike was only a Category 4 storm. Imagine the damage that a Category 6 hurricane would cause.

Better Prediction

New and improved technology is helping scientists predict hurricanes quicker and more accurately. One example of improved technology is the group of small satellites launched by US space scientists in 2016. They are called the Cyclone Global Navigation Satellite System (CYGNSS). They have sensors that can look inside hurricanes and collect a lot of very accurate data. They record ocean surface wind speeds in the eye and in the eye wall. This information is fed into supercomputers, which then create better 3-D animations of storms. They show the hurricane's shape and they can figure out its strength, direction, and how fast it is moving. They can predict possible storm surges, too. These new computer models can also show the effect of climate change, sea-level rises, and varying storm activity. This helps scientists improve accuracy and speed when forecasting hurricanes.

These are the paths of all known Category 5 Atlantic hurricanes between 1851 and 2014.

Where in the World?

The Great Hurricane of 1780 was the worst-ever Atlantic hurricane. This 300-mile- (483 km) wide storm appeared suddenly and unexpectedly. It went through the Lesser Antilles, a group of islands in the Caribbean Sea, and killed at least 24,000 people. Its exact strength was unknown because scientists did not then have the equipment to study hurricanes. However, its winds were so strong that the hurricane peeled the bark off trees. Only winds of more than 200 miles per hour (322 kph) could do that!

Disaster Zone

New York City

A lot of people live in New York City, and four of its five boroughs lie on three islands at the mouth of the Hudson River. It is one of the cities thought to be most at risk from future hurricanes and storm surges. Climate change is warming the oceans and fueling stronger hurricanes. Could the city soon face a mega hurricane?

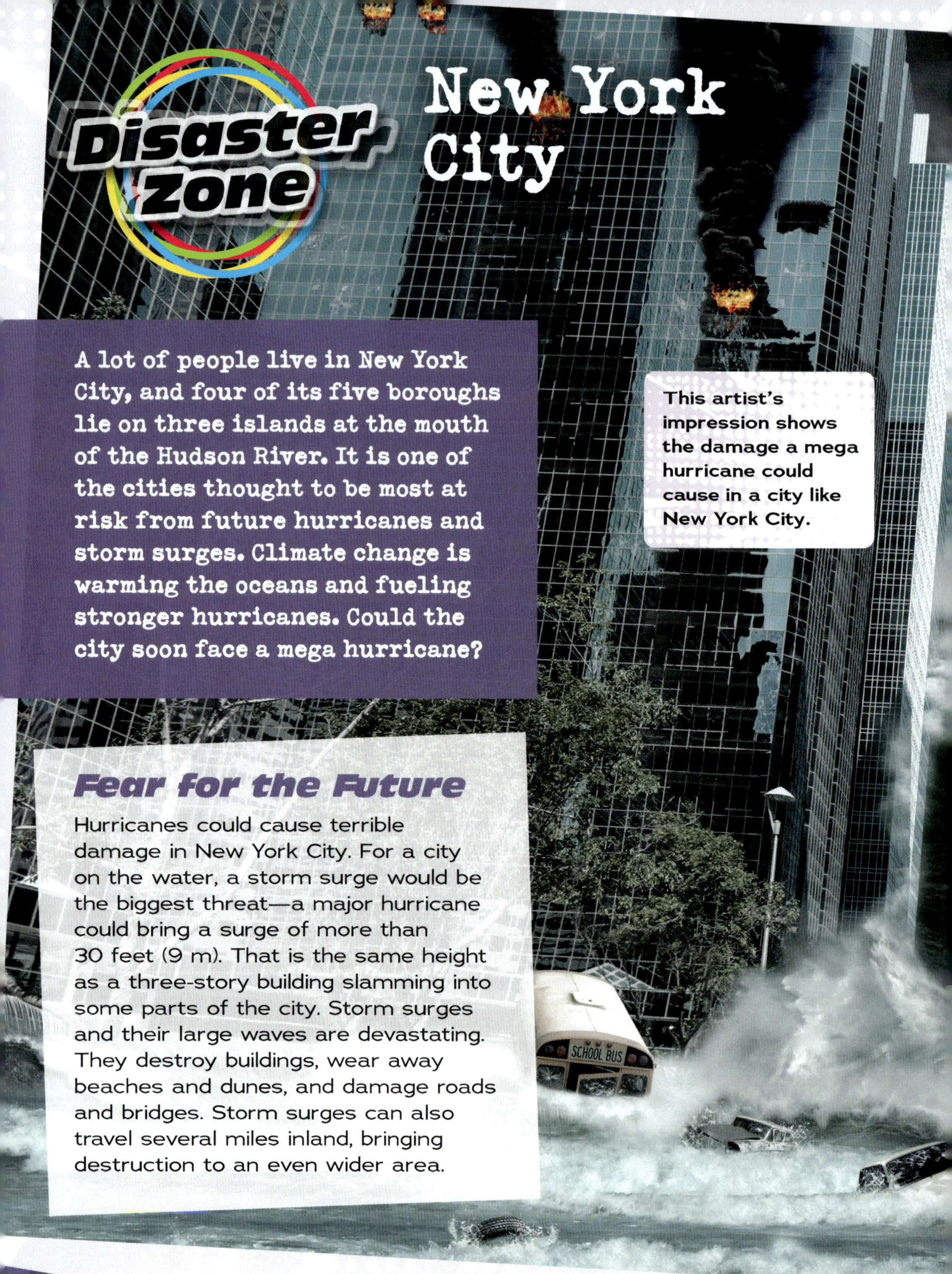

This artist's impression shows the damage a mega hurricane could cause in a city like New York City.

Fear for the Future

Hurricanes could cause terrible damage in New York City. For a city on the water, a storm surge would be the biggest threat—a major hurricane could bring a surge of more than 30 feet (9 m). That is the same height as a three-story building slamming into some parts of the city. Storm surges and their large waves are devastating. They destroy buildings, wear away beaches and dunes, and damage roads and bridges. Storm surges can also travel several miles inland, bringing destruction to an even wider area.

Increasing Risk

Today, more than 430,000 people are at risk of coastal flooding in New York City. The city has 100 square miles (260 sq km) of land lying in the coastal floodplain. A floodplain is an area of low land at risk from flooding. Since the start of the twentieth century, oceans have risen about 1 foot (30 cm) in New York Harbor. If sea levels continue to rise, by 2050, there will be 150 square miles (390 sq km) of land in the coastal floodplain. That means more people could be at serious risk from floods.

HURRICANE SANDY

In October 2012, Hurricane Sandy caused a huge storm surge almost 14 feet (4 m) high in New York Harbor. Waves raced through the city. They flooded streets, tunnels, and subways. Across New York and New Jersey, the storm left 60 people dead. 650,000 homes were hit, more than $70 billion of damage was caused, and 5.2 billion gallons (19.7 billion l) of sewage spilled into streets. Roads, bridges, and subways were closed.

The waves caused by Hurricane Sandy shorted out electrical transmission lines, plunging New York City into darkness.

Chapter 6

BE PREPARED

No one knows what lies ahead, but the evidence suggests that we should be ready to face more hurricane horrors in the future. Even the best forecasts may not be totally accurate because unpredictable storms can suddenly change in intensity. It is more important than ever to take precautions and plan for the worst.

Storm Shelters

Storm shelters are special buildings that are designed to keep people safe during a hurricane. Since Hurricane Sandy, New York City has made plans for up to 120,000 emergency shelters. Public storm shelters are large buildings where many people can go. In more **remote** places, people often buy or build their own shelters near their homes. An example of a hurricane shelter is one made from steel panels welded strongly together into a box-shape. Some are made from concrete. Shelters have strong steel doors. They are bolted to concrete bases so they can stand up to strong winds and flying debris. Some people have rooms inside their houses, usually basements, that are strengthened for protection.

Where in the World?

In some places, there are sea walls to protect coastal towns from dangerous storm surges. A hurricane hit Galveston, Texas, in 1900, killing 6,000 people. Afterward, a huge sea wall was built to stop storm waves reaching the shore. Thankfully, as a result Galveston even survived Hurricane Ike in 2008.

Making Structures Safer

Even if people escape to shelters, they need to protect their homes during a storm. They may put up strong storm panels or shutters to cover windows to stop hurricane winds from smashing the glass. They cut off dead branches and cut down any old trees that hurricane winds may bring down. People also fit metal straps or clips to attach their roofs to exterior walls. Hurricane-**prone** cities may have storm surge drains. The large drains allow a lot of water to be taken away from the streets to prevent floods.

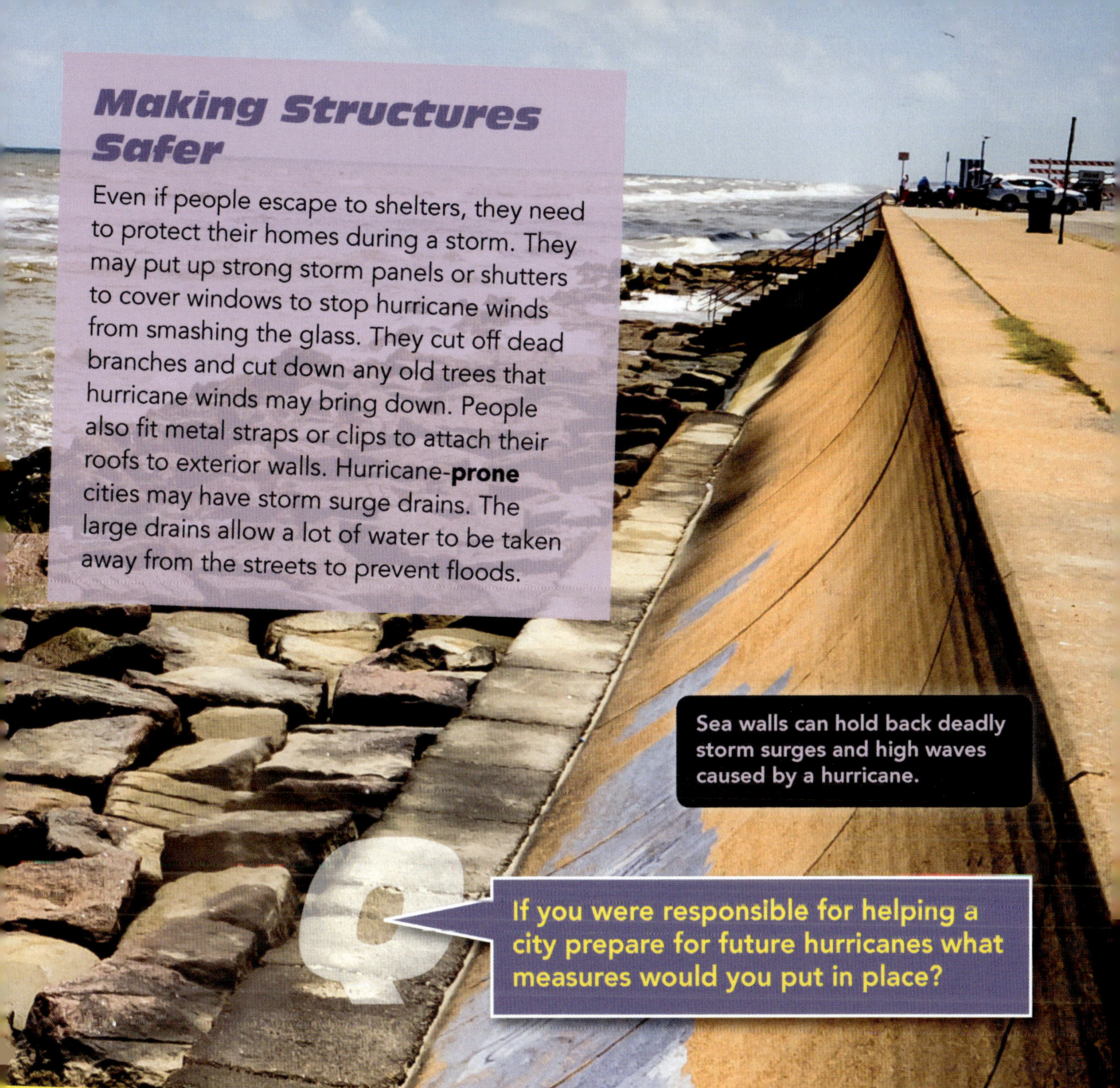

Sea walls can hold back deadly storm surges and high waves caused by a hurricane.

Q If you were responsible for helping a city prepare for future hurricanes what measures would you put in place?

Galveston lies directly on the coast and is vulnerable to the waves that batter it.

HURRICANE WARNING SYSTEMS

Weather experts try to give people as much warning as possible so they can leave an area before a hurricane arrives. However, to be certain of survival, people need to be ready. They must make emergency plans long before any warnings are given.

Learning about hurricanes helps people understand why they should prepare and plan for these possible disasters.

Make a Plan

Every household in a hurricane zone needs a plan, and each family member must know what to do. They need to know how to find each other, and how to get in touch if a hurricane strikes. Every building with people should have a plan. This plan tells people where their nearest shelter is and what to take with them. Everyone needs to know the fastest route to get to the shelter. They must know how to get there, for example, if they do not have a car.

Everyone should have an emergency kit. An emergency kit is a box or bag that contains essential items. These are the things that can help people while they are in a shelter or if they have to leave. It might include cans of food, long-life drinks, and a supply of fresh water. Other important things to pack are clothes and bedding, a first aid kit, and medicines. A flashlight, batteries, and a charged cell phone are also important. Papers such as identity documents must be packed, too.

Watching for News

During a hurricane season, everyone should watch the news or listen to the radio. People can check online forecasts to find out more information about what's going on. A hurricane watch is given when there is a chance that hurricane conditions might arrive within 48 hours. A hurricane warning is given when winds of at least 74 miles per hour (119 kph) are expected within 36 hours. When there is a hurricane alert, local authorities will tell people if there is time to leave the area or if they should go to storm shelters.

During hurricane season people should be ready to leave their homes at a moment's notice.

WARNING TIMES

Hurricane forecasts change all the time. Five days before a likely hurricane, forecasts can be up to 350 miles (563 km) out. Twenty-four hours before landfall, this drops to 100 miles (161 km). This can make the difference between a government ordering people to leave or stay—and the difference between life and death.

SURVIVING A HURRICANE HORROR

Hurricanes are forces of nature that cannot be stopped. While we cannot stop hurricanes from happening, science, technology, and being prepared can help most people survive these terrifying storms.

Knowledge Saves Lives

Understanding hurricanes can keep people alive. Learning about hurricanes and the warning systems mean people know what to do and when. For example, the eye of the hurricane is the calmest part of the storm. People who know that will not leave their homes or shelters even when it seems that the storm has passed over. They know that they may be in the eye of the storm, and that it is not yet over. They will wait until the violent winds from the opposite side of the eye wall have blown through. Also, when people understand how hurricanes cause storm surges, they will do everything to stay safe.

Hurricane warnings are made as clear as possible.

Where in the World?

In 2018, Hurricane Michael headed toward land. It was fueled by the warm waters of the Gulf of Mexico. With winds that roared like a jet engine, Michael tore apart buildings and snapped power lines in half. Before Michael hit Florida, more than half a million people were ordered or advised to leave. As a result, just two people lost their lives.

People Who Help

There are also many people who help as soon as a hurricane warning is given. They are the rescue workers. They use boats, helicopters, and other vehicles to rescue people during a hurricane. People may be stranded or injured by winds, floods, and storm surges. Rescuers help people get to safety. The may go to the hospital or to a storm shelter. During the worst disasters, thousands of emergency service workers, troops, and volunteers work together to keep people safe.

Scientists are working hard to find ways of creating better forecasts. This will help them give earlier warnings, and that will give people more time to get to safety. Although hurricanes are dangerous, if people are careful and listen to and follow warnings when they are given, they can keep themselves safe.

Helicopters lift people to safety when they are caught up in a hurricane.

Glossary

air pressure a pushing force caused by the weight of air above
atmosphere the blanket of gases surrounding Earth
axis an imaginary line through Earth's poles
climate change a change in the pattern of the world's weather
clockwise the direction in which the clock hands normally move
counterclockwise the opposite direction to which the clock hands normally move
crops plants grown for food or other purposes
dam a barrier built to hold back water
data information collected for study
debris pieces from something that has been destroyed
equator an imaginary horizontal line around the center of Earth
evacuate to move from danger in an organized way
flash floods sudden floods caused by large amounts of rainfall
flooded covered normally dry land in water
forecast to predict future weather
fossil fuels oil, coal, and gas, formed from long-dead plants and animals
Global Positioning System (GPS) a satellite system used to find directions and locations
hemisphere half of Earth
humidity a measure of the amount of water vapor in the air
landfall arrival on land
orbit the curved path through which objects in space move around a planet or star
parallel alongside or in the same direction as something else, while keeping the same distance apart and never crossing paths
polluted containing harmful substances
predicting telling what might happen in the future
prone likely to do, get, or suffer from something
radar a system that uses radio waves to locate and measure objects at a distance
remote far away from towns and cities
satellite a human-made object that orbits Earth
sensors devices that are used to record properties such as heat, sound, and light
sound barrier a sudden large increase in drag caused when an airplane approaches
storm surges bulges in the sea level caused by a storm on the ocean
submerged completely covered with water
thermal image an image that uses colors to show the differences in the amount of heat in an object
transmission towers tall structures that carry power lines
transmit to send out
tropical describes the hot region near the Earth's equator
water vapor very fine droplets of water suspended in the air

Find Out More

Books

Finch, Fletcher C. *Hurricane Katrina* (History Just Before You Were Born). Gareth Stevens Publishing, 2019.

Haney, Johannah. *Natural Disasters! With 25 Science Projects for Kids.* Nomad Press, 2020.

Lew, Kristi. *Hurricanes* (Nature's Mysteries). Britannica Educational Publishing, 2019.

Sillett, Julia. *Hurricanes Harvey, Irma, Maria, and Nate* (Disaster Alert!). Crabtree Publishing, 2018.

Websites

Educate yourself about hurricanes at:
https://eo.ucar.edu/kids/dangerwx/hurricane3.htm

Find out more about how hurricanes work at:
https://science.howstuffworks.com/nature/natural-disasters/hurricane.htm

There is more hurricane information at:
https://scijinks.gov/hurricane

For more information from NASA about how hurricanes form go to:
https://spaceplace.nasa.gov/hurricanes/en

Publisher's note to educators and parents:
All the websites featured above have been carefully reviewed to ensure that they are suitable for students. However, many websites change often, and we cannot guarantee that a site's future contents will continue to meet our high standards of educational value. Please be advised that students should be closely monitored whenever they access the Internet.

Index

About the Author

Louise Spilsbury is an award-winning author who has written hundreds of wonderful books for children. She has written many books about natural disasters and has a healthy respect for these awesome forces of nature!